S0-BII-552

Understanding the Elements of the Periodic Table™

ZINC

Kristi Lew

rosen publishing's
rosen central

New York

Published in 2008 by The Rosen Publishing Group, Inc.
29 East 21st Street, New York, NY 10010

First Edition

Library of Congress Cataloging-in-Publication Data

Lew, Kristi.
Zinc / Kristi Lew.—1st ed.
p. cm.—(Understanding elements of the periodic table)
Includes bibliographical references and index.
ISBN-13: 978-1-4042-1407-1 (library binding)
1. Zinc. 2. Periodic law. 3. Chemical elements. I. Title.
QD181.Z6L49 2008
546'.661—dc22

2007027125

Manufactured in Malaysia

On the cover: Zinc's square on the periodic table of elements; *(inset)* model of the subatomic structure of a zinc atom.

Contents

Introduction

Have you ever noticed people at the beach with white goop all over their noses and lips? If so, you have seen a product that contains one of the most common zinc (chemical symbol: Zn) compounds—zinc oxide (ZnO). People use a paste containing zinc oxide to block the harmful rays of the sun, which can cause a nasty burn. Zinc is sometimes called "the Great Protector." Not only does zinc protect people from the sun, but it also protects iron and steel from corroding, or rusting. That is a pretty big job when you think about all the buildings, bridges, cars, and other things made of steel that we rely on every day.

Zinc is found in hundreds of household products—from cereals and vitamins to soaps and shampoos. Zinc compounds (zinc combined with other elements) are used to produce white and yellow paints, ceramics, rubber, wood preservative, dyes, and fertilizers. Zinc compounds are also used by drug companies as ingredients in sunblock, diaper rash ointment, deodorants, antidandruff shampoos, and treatments for athlete's foot, acne, and poison ivy.

Zinc is also important for good health. The mineral is needed for healthy skin and bones. It helps keep your sense of smell and taste functioning properly. And it plays an important part in a healthy immune system.

Today, the so-called copper penny is actually mostly zinc. When it was introduced in 1787, the penny was made entirely of copper. Sometimes,

"Copper" pennies are made mostly of zinc. A thin layer of copper coats only the outside of the coins.

however, the rising value of copper leads the U.S. Mint to use less of the metal in its pennies. During World War II, for example, copper was vitally important to the American war effort. So, in 1943, the U.S. Mint pressed a batch of pennies using mostly low-grade steel. After the war, pennies were once again made from copper, but with a little zinc. When the value of copper rose in the 1980s, the mint decided to make its pennies mostly from zinc. Today, the copper content of the penny is a mere 2.5 percent, with the rest of the coin (97.5 percent) being zinc.

As you can see, zinc is an important and useful substance. It is no wonder Americans use more than one million tons of zinc a year!

Chapter One
From Ore to Metal

Pure zinc is a bluish white, solid metal. It exists naturally in rocks and soil, and small amounts are also found in air, water, plants, and animals.

History of Zinc

There is no record of when zinc was first discovered. It is believed zinc minerals were used in China and India during medieval times. Minerals are naturally occurring substances made through geological processes, like volcanic eruptions or the evaporation of ancient seas. Minerals may be made up of only one element or be a combination of many different elements. Zinc minerals have been mined for a very long time.

Zinc alloys have also been used for centuries. Alloys are mixtures of two or more metals. They can be made by melting the metals and mixing them together. The oldest known object made of a zinc alloy was a religious idol found in prehistoric ruins in eastern Europe. Objects made of brass, a mixture of zinc and the metal copper, have also been discovered, some dating back two thousand years.

From the thirteenth through the sixteenth centuries, metallic zinc and zinc oxide were produced in Zawar, India. The metal was used to make brass, and the zinc oxide was used as a medicine. To get pure zinc,

Zinc is a bluish white, shiny solid. A coating of dull-grey zinc carbonate forms when the metal is exposed to oxygen and carbon dioxide in the air.

mineral rocks containing zinc were chemically reacted with wool or other natural materials containing carbon (C).

Humans have used metals like zinc for thousands of years. However, most of our knowledge regarding the chemistry of metals and other substances did not come about until the eighteenth century. German scientist Andreas Sigismund Marggraf (1709–1782) is generally credited with identifying zinc as an element in 1746. Like the people in India, he used a zinc mineral to get relatively pure metallic zinc. The English name of the element is taken from its German name, *zink*.

In 1746, German chemist Andreas Sigismund Marggraf *(left)* isolated zinc as an element.

Where Is Zinc Found?

An ore is a mineral deposit containing a valuable metal or other resource. Earth's crust has about two billion tons of zinc ore available for mining. About 80 percent of the world's zinc mines are tunneled underground. Another 8 percent of zinc deposits are closer to the surface, though, and open-pit mining is more common in these areas. The remaining 12 percent of zinc-mining operations are a combination of underground and open-pit mines.

Zinc is mined in more than fifty countries. The leading producer of zinc metal is China, followed by Australia, Peru, Canada, and the United States. In the United States, Alaska produces the most zinc. But the metal is also mined in Tennessee and Missouri.

How Does Ore Become Metal?

Zinc sulfide (ZnS), also known as sphalerite or zinc blende, is the most common zinc mineral. Ninety-five percent of all the zinc mined around the world today is naturally found as zinc sulfide.

To get pure zinc metal from zinc sulfide, the sulfur must be removed. Zinc blende is heated in large furnaces to greater than 1,650° Fahrenheit

For open-pit mining, surface rock and soil are removed to expose the ore below. This Australian open-pit mine produces copper, silver, and lead, as well as zinc.

(900° Celsius). At such high temperatures, the zinc sulfide reacts with oxygen (O) in the air. This produces zinc oxide, which is a solid, and sulfur dioxide (SO_2), which is a gas. Zinc metal is extracted from the zinc oxide by heating the oxide with carbon. The carbon reacts with zinc oxide, forming carbon dioxide (CO_2) and leaving zinc metal.

The Red Dog Mine

The Red Dog Mine is an open-pit mine located in Alaska, 85 miles (137 kilometers) north of the Arctic Circle. It is the largest zinc mine in the world, producing 10 percent of the world's supply. Scientists believe the zinc in

Sphalerite, or zinc sulfide, is the most common zinc mineral. In addition to zinc, some sphalerite deposits yield useful amounts of cadmium (Cd), gallium (Ga), and indium (In).

Red Dog formed about 330 million years ago. Zinc and other metals were present in hot fluids below an ancient sea floor in the Red Dog Basin. As these fluids moved up from Earth's core to the floor of the seabed, they cooled and formed a rock called black shale. After hundreds of thousands of years, erosion exposed the mineral deposit, which is now at the surface.

Renewable Resource

Zinc can be recycled over and over without affecting its chemical or physical properties. About 30 percent of the world's zinc metal comes from recycled material. In the United States, recycling produces about 400,000 tons (362,874 metric tons) of zinc a year.

Other Zinc Minerals

Zinc sulfide is not the only zinc mineral. Others include zinc carbonate, zinc oxide, and hydrated zinc silicate. Zinc carbonate is also called smithsonite. Smithsonite is sometimes used to produce zinc metal, but mostly it is polished and used as a decorative stone. The common name for zinc oxide is zincite, and hydrated zinc silicate is called hemimorphite. Both of these minerals are important zinc ores. There is also a natural mixture of zinc, manganese (Mn), and iron oxide called franklinite. Franklinite was once an important zinc ore, but it has been found in large amounts only in Franklin, New Jersey. The last zinc mine in Franklin closed down in 1986, so franklinite is no longer used as an ore.

Chapter Two
Zinc Atoms and the Periodic Table

Zinc and all of the other elements in the periodic table are made of atoms. In this chapter, we'll take a closer look at the structure of atoms. We'll also see how the special structure of a zinc atom makes it unique among the elements.

What Are Atoms and Elements?

Atoms are the extremely tiny building blocks of all matter. Desks, chairs, air, water, and anything else that can be touched is made up of atoms. Atoms, in turn, are made up of even tinier subatomic particles called protons, neutrons, and electrons. Protons and neutrons are found in the nucleus, or central core, of the atom. Electrons circulate rapidly around the nucleus in electron shells. Electrons in different shells are circulating at different energy levels.

Elements are substances made up of one type of atom. The individual protons, neutrons, and electrons that make up all elements are the same. However, elements differ from one another because of the unique number and arrangement of their subatomic particles. The number of protons in an atom, called the atomic number, determines the element. Zinc, with atomic number 30, has atoms with thirty protons. If an atom has twenty-nine protons, it is not an atom of zinc. In fact, it is an atom of copper.

Protons (yellow) and neutrons (black) make up the nucleus of an atom. Electrons (light blue) travel in shells outside the nucleus. The diagram at right shows the structure of a carbon atom.

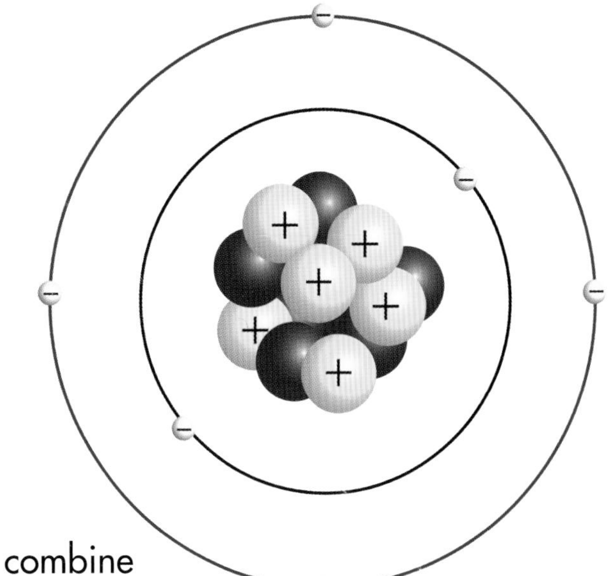

What Are Compounds?

When two or more elements chemically combine with one another, the result is called a chemical compound. The most common zinc compound is zinc oxide. Zinc oxide is formed when the elements zinc and oxygen chemically react with each other. The reaction creates chemical bonds that hold the zinc and oxygen atoms together. Zinc oxide and other compounds of zinc occur naturally in Earth's crust. In fact, zinc reacts so readily with other elements that the pure form of the element is rarely found in nature. You'll read more about zinc compounds in chapter four.

Elements and the Periodic Table: A Brief History

Scientists have identified more than 100 elements. To organize the elements, chemists use a chart called the periodic table. (See pages 38–39.) This table arranges the elements so that ones with similar chemical and physical properties are grouped together in rows and columns. Color, smell, and state of matter (liquid, solid, or gas) at room temperature are some examples of physical properties. Chemical properties describe how an element behaves chemically. For example, how an element bonds and whether an element will burn are chemical properties.

The periodic table we use today is based on one developed by Dmitry Mendeleyev in the late 1860s. Mendeleyev was a Russian chemistry professor. He was convinced the sixty-three elements known at the time showed regular (or "periodic") patterns in their properties. Mendeleyev arranged the elements, in order of increasing atomic mass, into groups with similar properties. He left gaps in his periodic table in places where none of the elements he was working with fit neatly into a group. He correctly believed that one day, scientists would identify elements with the right properties to fit into the blank spaces. As of 2007, scientists had discovered or created a total of 117 elements, a few of which are not yet accepted by all authorities.

The Periodic Table: Periods

Elements are now arranged in the periodic table by increasing atomic number, or number of protons in the atomic nucleus, instead of atomic mass. The table features seven horizontal rows, which are called periods. The period number of an element represents the number of electron shells or energy levels of an atom of that element. Zinc, in period 4, has atoms with four electron shells.

The Periodic Table: Groups

The vertical columns of the periodic table are called groups or families. Elements in the same group share some properties. There are eighteen groups on the modern periodic table, numbered one through eighteen. Sometimes, periodic tables also show an older group numbering system. In the older system, groups are numbered with Roman numerals and letters: IA through VIIA, IB through VIIIB, and O. Zinc is found in group 12 (IIB).

The periodic table is often divided into four parts: main group elements, transition metals, lanthanides, and actinides. The main group elements are in groups 1 and 2 (IA and IIA), and then groups 13 through 18 (IIIA through VIIA, and O). The thirty-eight elements in groups 3 through 12 (IB through VIIIB), of which zinc is one, are called the transition metals. The lanthanides fit into period 6, and the actinides in period 7. The lanthanides and actinides are usually found as two separate rows at the bottom of the periodic table. This arrangement keeps the periodic table more compact and easier to read.

Zinc and Its Electrons

Elements in the same group on the periodic table have similar physical and chemical properties because they have electrons arranged in similar patterns. The electrons moving around the atomic nucleus are negatively charged. The protons, concentrated in the center of the atom, in the nucleus, are positively charged. An atom has the same number of protons as electrons, making the negative and positive charges balance out. Such atoms are electrically neutral. Zinc, with atomic number 30, has thirty protons. A neutral zinc atom, therefore, will also have thirty electrons.

The way electrons are arranged in their shells around the atom's nucleus is called the electron configuration. In all atoms, the first shell can hold up to two electrons, and the second shell can hold up to eight electrons. The third shell can hold up to thirty-two electrons, and the higher shells can hold even more. Zinc, with thirty electrons, has two electrons in its first shell, eight in the second, eighteen in the third, and two electrons in the fourth shell.

The two electrons in zinc's fourth shell are called valence electrons. Valence electrons are the electrons in an element's outermost shell, or highest energy level. Electrons in the outermost level can be gained, lost, or shared during chemical bonding to make chemical compounds. Cadmium (Cd) and mercury (Hg), the other elements in zinc's group 12 on the periodic table, also have two valence electrons. The similar electron

Group

Period	VIIIB 10	IB 11	IIB 12	IIIA 13	IVA 14	VA 15	VIA 16	VIIA 17	O 18
1									2 4 He Helium
2				5 11 B Boron	6 12 C Carbon	7 14 N Nitrogen	8 16 O Oxygen	9 19 F Fluorine	10 20 Ne Neon
3				13 27 Al Aluminum	14 28 Si Silicon	15 31 P Phosphorus	16 32 S Sulfur	17 35 Cl Chlorine	18 40 Ar Argon
4	28 59 Ni Nickel	29 64 Cu Copper	30 65 Zn Zinc	31 70 Ga Gallium	32 73 Ge Germanium	33 75 As Arsenic	34 79 Se Selenium	35 80 Br Bromine	36 84 Kr Krypton
5	46 106 Pd Palladium	47 108 Ag Silver	48 112 Cd Cadmium	49 115 In Indium	50 119 Sn Tin	51 122 Sb Antimony	52 128 Te Tellurium	53 127 I Iodine	54 131 Xe Xenon
6	78 195 Pt Platinum	79 197 Au Gold	80 201 Hg Mercury	81 204 Tl Thallium	82 207 Pb Lead	83 209 Bi Bismuth	84 209 Po Polonium	85 210 At Astatine	86 222 Rn Radon
7	110 271 Ds Darmstadtium	111 272 Rg Roentgenium	112 277 Uub Ununbium		114 289 Uuq Ununquadium		116 289 Uuh Ununhexium		

65 159 Tb Terbium	66 162 Dy Dysprosium	67 165 Ho Holmium	68 167 Er Erbium	69 169 Tm Thulium	70 173 Yb Ytterbium	71 175 Lu Lutetium
97 247 Bk Berkelium	98 251 Cf Californium	99 252 Es Einsteinium	100 257 Fm Fermium	101 258 Md Mendelevium	102 259 No Nobelium	103 262 Lr Lawrencium

Based on its physical and chemical properties, zinc belongs in period 4 and group 12 (IIB) of the modern periodic table. It is a transition metal.

configuration among group 12 elements is the reason they share some chemical properties.

Group 12 Elements

Zinc, cadmium, and mercury are all lustrous, or shiny, metallic elements. They conduct electricity fairly well, and they mix easily with other metals. Both zinc and cadmium are relatively soft metals that are ductile and malleable. A metal is malleable if it can be hammered or rolled into different shapes without breaking. Ductile means the metal can be stretched into wires.

Ununbium (Uub), the other element in the group containing zinc, cadmium, and mercury, was made in a laboratory for the first time on February 9, 1996. It does not occur in nature, and scientists have made only a few atoms of it, so not much is known about it yet.

Zinc's Bonding Behavior

To create chemical compounds with other elements, atoms can lose, gain, or even share electrons. Zinc gives up the two valence electrons in its fourth electron shell. The fourth

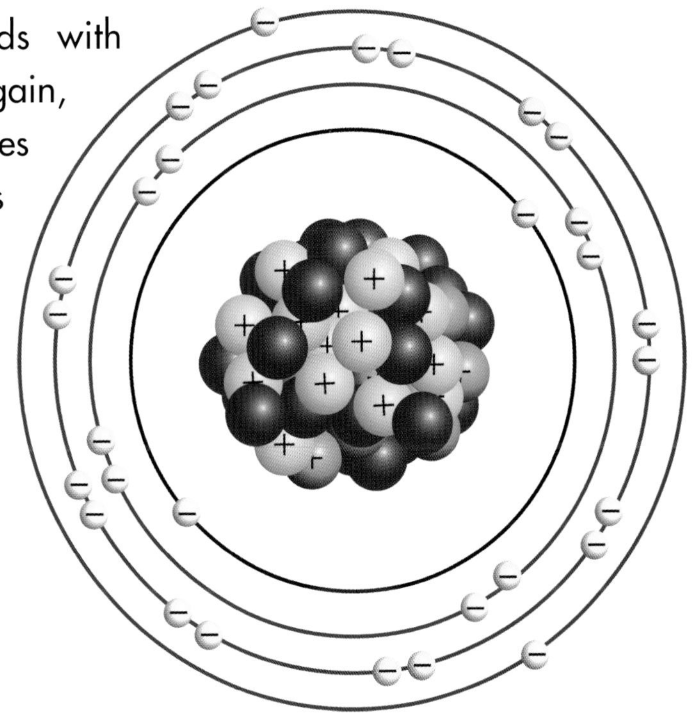

At right is a model of a zinc atom. The nucleus holds thirty positively charged protons (green). Thirty negatively charged electrons (light blue) move around the nucleus, in four separate shells. The black spheres in the nucleus represent the atom's neutrons.

Zinc  Snapshot

Chemical Symbol:	Zn
Classification:	Transition metal
Properties:	Shiny, bluish white metal; ductile; malleable
Discovered by:	Known since ancient times; isolation and identification as an element credited to Andreas Marggraf in 1746
Atomic Number:	30
Atomic Weight:	65.41 atomic mass units (amu)
Protons:	30
Electrons:	30
Neutrons:	34, 36, 37, 38, 40 (approximate average, 35)
State of Matter at 68° Fahrenheit (20° Celsius):	solid
Density at 68°F (20°C):	7.134 g/cm^3
Melting Point:	787.15°F (419.53°C)
Boiling Point:	1,664.6°F (907.0°C)
Commonly Found:	In Earth's crust; trace amounts found in air, water, plants, and animals

shell disappears, and zinc is left with three filled electron shells. This is an example of a general principle in chemistry: when atoms react, they lose or gain electrons to achieve a filled outer energy level.

With the loss of its two valence electrons, the zinc atom ends up with two more protons than it has electrons. The positive charges of the protons give the zinc particle a positive charge. Since two negatively charged electrons

Isotopes of Zinc

In addition to protons and electrons, there is another type of subatomic particle called a neutron. Neutrons are neutral, which means they do not have an electrical charge, but they do have mass. The mass of a neutron is almost the same as the mass of a proton. The mass of an electron, however, is a tiny fraction (1/2,000) of the mass of a proton or neutron. As a result, the protons and neutrons in the nucleus of an atom make up nearly all of the atom's mass.

Some elements have special forms, called isotopes. Different isotopes of an element have atoms with different numbers of neutrons but the same number of protons. Zinc has five natural isotopes: zinc-64, zinc-66, zinc-67, zinc-68, and zinc-70. (The number written behind the element's name represents its approximate mass.) The most commonly occurring isotope of zinc is zinc-64. The standard atomic mass of an element listed on the periodic table is actually an average of the weights of the element's isotopes. Just as the average weight of a group of people does not correspond to the weight of any one of them, the atomic weight of an element is not necessarily the weight of any atom. Zinc's average atomic weight is 65 atomic mass units (amu). So, on average, zinc atoms have thirty-five neutrons.

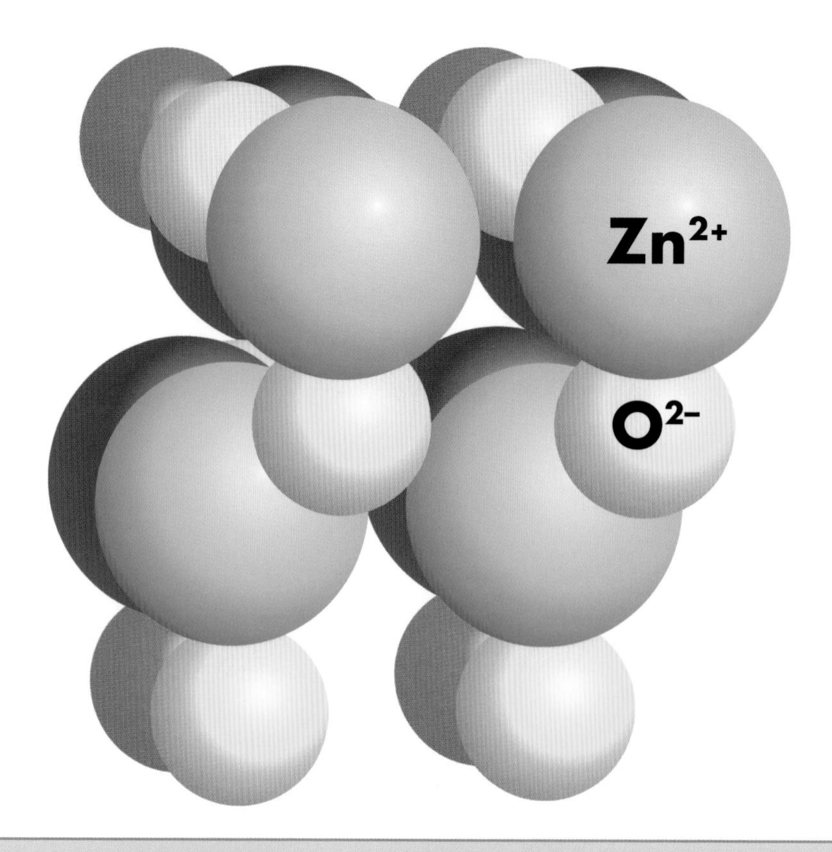

Above is an example of ionic bonding. Positively charged zinc ions (green) combine with negatively charged oxygen ions (yellow) to form a molecule of zinc oxide.

were lost, the particle's charge is a positive two. This charged particle is no longer a normal zinc atom but a zinc ion. Positive zinc ions combine with negative ions in an ionic bond. Zinc easily forms ionic bonds with chlorine (Cl), oxygen, sulfur (S), and other nonmetals to form ionic zinc compounds.

Chapter Three
The Great Protector

A chemical reaction between metal and air or water in the atmosphere can cause the metal to break down. This destructive process is called corrosion. When iron (Fe) is exposed to air, for example, it forms iron oxide (Fe_2O_3), better known as rust. However, when zinc is exposed to oxygen in the air, it quickly forms a hard zinc oxide coating. The zinc oxide reacts with carbon dioxide (CO_2) in the air to form zinc carbonate ($ZnCO_3$). The dull-grey zinc carbonate coating protects the rest of the metal from further corrosion.

Galvanizing

Zinc is the fourth most commonly used metal in the world behind iron, aluminum (Al), and copper. Worldwide, more than seven million tons of zinc are produced every year. More than 50 percent of that zinc is used for galvanizing steel.

Galvanization is the process of coating a metal with zinc. One way to galvanize steel is to dip the steel into molten, or melted, zinc. This is called hot-dip galvanization. Introduced in France in 1836, hot-dip galvanization is the oldest anticorrosion process in the world.

Another way to deposit a zinc coating onto steel is through a process called electroplating. This process takes advantage of a zinc

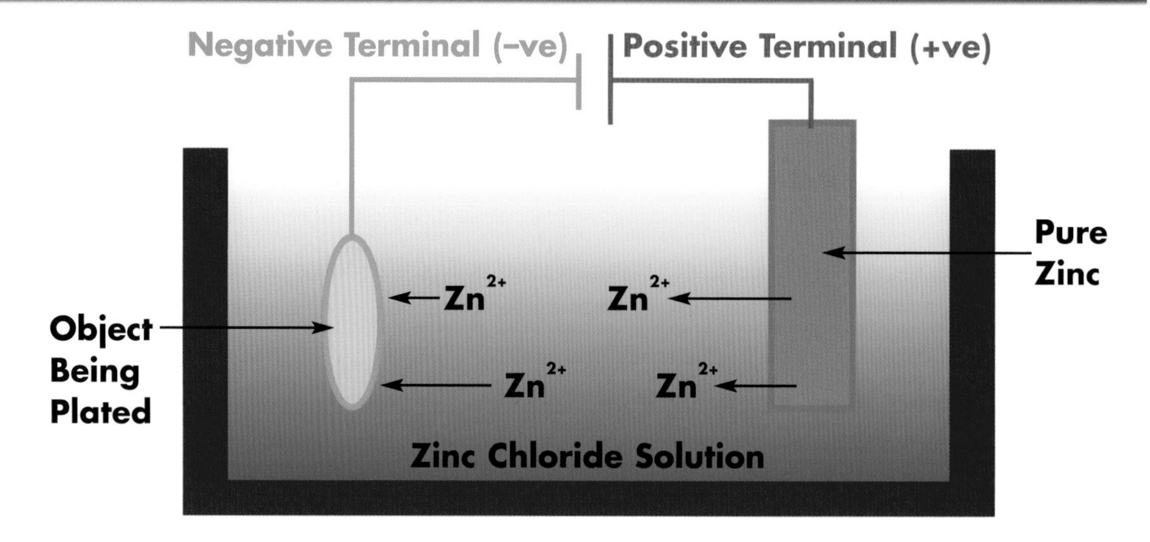

Shown here is a diagram of the electroplating process. With the help of electricity, zinc ions (Zn^{2+}) in the zinc chloride solution are attracted to the negatively charged object being plated. The ions then become metallic zinc atoms, coating the object with zinc metal. Pure zinc supplies additional zinc ions to continue the process.

atom's electrical properties. To electroplate a piece of steel with zinc, the part is first cleaned and placed in a plating tank. In the tank, a wire connected to the negative terminal of a power source is attached to the steel part. This causes the steel part to develop a negative charge and become a negative electrode. A zinc rod is placed into the tank and attached to the positive pole of the power source. The zinc rod is the positive electrode.

The tank is then filled with a solution containing a salt, or ionic compound, of zinc. Zinc chloride ($ZnCl_2$) is an example of a zinc salt commonly used in electroplating. When zinc chloride is dissolved in water, two types of ions form: zinc ions with a positive charge and chloride ions with a negative charge.

Because the steel part is negatively charged, it attracts the positively charged zinc ions in the salt solution. When the zinc ions reach the negative electrode, electrons flowing from the power source through the steel combine with the zinc ions, forming zinc atoms. The addition of electrons is called reduction. So, the zinc ions are reduced to metallic zinc atoms, which form a coating of zinc metal on the steel part.

At the positive electrode, electrons are being removed from the zinc rod by the power source. The removal of electrons is called oxidation. When electrons are removed from the zinc atoms, the atoms turn into zinc ions, which dissolve in the solution. The oxidation process replaces the zinc ions taken out of the solution at the negative electrode as zinc atoms are deposited onto the steel part.

The coatings from both hot-dip galvanization and electroplating prevent oxygen and moisture from corroding the steel underneath. Galvanized steel is used for car bodies, fencing, bridges, highway guardrails, outdoor garbage cans, and many other useful things. Even if the zinc coating is scratched, the zinc will still protect the steel because zinc is more easily corroded than the iron in the steel.

Sacrificial Anodes

Sometimes, a piece of iron cannot be plated with zinc for some reason. When this is the case, the iron item can still be protected by a piece of zinc attached to it. For example, a piece of zinc metal can be attached to the iron rudder of a ship. The zinc reacts with elements in the water, causing it to slowly corrode and disappear. If the zinc were not there, the iron in the rudder of the ship would corrode instead. Zinc used this way is called a sacrificial anode. Once it is gone, it is replaced with another piece of zinc, so the iron rudder is always protected.

Zinc Alloys

An alloy is a mixture of two or more metals. Alloys have different properties than the pure elements they are made from. There are several useful alloys made with zinc.

Brass is a mixture of zinc and copper. It starts out a bright yellow color, but as it ages and reacts with other chemicals in the air, it produces an outer coating called a patina. This outer coating is corrosion, but it protects the rest of the metal from further damage. Polishing off the patina will return the original shine to the brass.

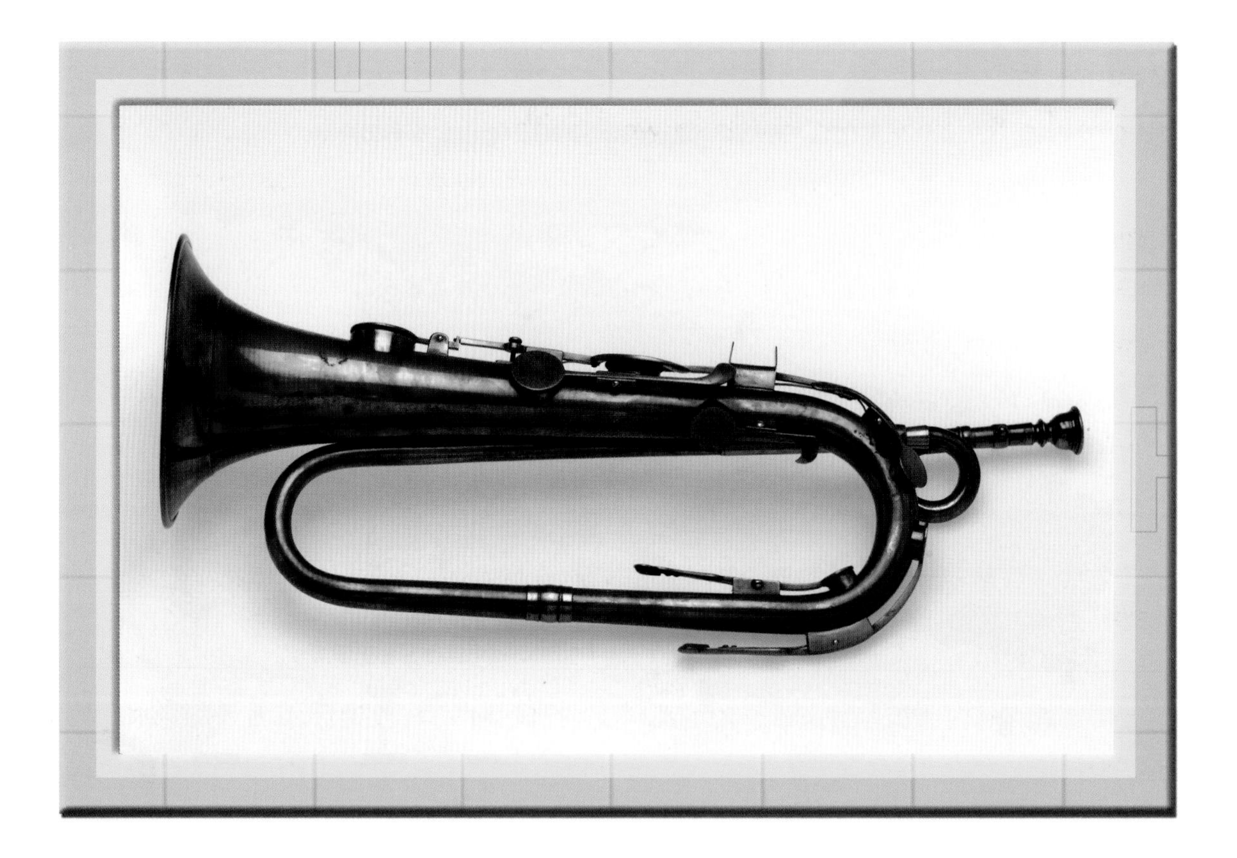

Brass, a mixture of zinc and copper, is a useful alloy. It is often used to make musical instruments, such as this antique bugle.

Zinc in Fireworks

Zinc is used to make the smoke and flash in fireworks. Sparklers can be made of zinc, magnesium, aluminum, steel, or iron dust, too. The dust burns quickly and at high temperatures, making the bright, shimmering sparks.

Brass has been around for at least 1,500 years. Ancient Romans used the alloy to make coins, kettles, and decorative items. Brass is still used extensively to make musical instruments, decorative outdoor lights, screws, and other hardware that needs to be resistant to corrosion. Most brass alloys contain between 20 and 50 percent zinc. Higher percentages of zinc make the brass too brittle to work with.

If tin (Sn) is added to a mixture of zinc and copper, an alloy called bronze is formed. Bronze has many of the same applications as brass. Both brass and bronze are harder and stronger than the pure elements they are made from.

Zinc can also be alloyed with lead (Pb) and tin to make solder. Solder is melted and used to join pipes, electrical components, and other metallic items. Zinc mixed with aluminum in about a three-to-one ratio yields a strange alloy called Prestal. Prestal is unique because it is as strong as steel, but it can be molded as easily as plastic.

Zinc alloys are commonly used in die casting. To make a die cast, the alloy is melted, poured into a mold, and allowed to cool. The metal takes on the shape of the mold. Bathroom plumbing fixtures, door and window hardware, feet and wheels on office furniture, tools, automobile parts, and electronic components are just a few of the items die cast from zinc alloys.

Chapter Four
Zinc Compounds

A wide variety of everyday items contain zinc compounds. From paints, plastics, printing inks, and rubber products to cosmetics, pharmaceuticals, soaps, and batteries, zinc compounds are just about everywhere.

Zinc Oxide as Sunscreen

Zinc oxide is probably the best known and the most used compound of zinc. It is found as an ingredient in sunscreen and as a pigment (coloring agent) in plastics, cosmetics, copier paper, and wallpaper. Calamine lotion, a mixture of zinc and iron oxides, can greatly reduce the itching associated with poison ivy, poison oak, or poison sumac. It dries up the blisters and helps to prevent infection caused by scratching.

Zinc oxide is a white powder that does not dissolve easily in water. This is good news if it is being used as a sunscreen because it is less likely to be washed away by water or sweat. One advantage zinc oxide sunscreen has over others is that it blocks and reflects both UVA and UVB solar rays before they reach the surface of the skin. UVB solar rays are the ones mostly responsible for sunburns and skin cancers. But UVA rays penetrate deeper into the skin and cause premature aging associated with being

Sunscreens that contain zinc oxide block and reflect both UVA and UVB solar rays, reducing the risk of sunburn, skin cancer, and premature aging.

exposed to the sun for long periods. Studies have also shown UVA rays increase the cancer-causing effects of UVB rays.

Zinc Oxide and the Photocopy

Zinc oxide is photoconductive, which means it conducts electricity when it is exposed to light. This property makes zinc oxide useful in photocopiers. In a photocopier, a plate made of zinc oxide is charged with electricity. When light is reflected from the document to be copied, it hits the charged plate of zinc oxide. The areas of the document that do not have print on them will reflect light onto the plate, taking the electric charge away from those areas of the plate. When that happens, the only parts of the plate still electrically charged are the dark parts corresponding to the ink on the paper being copied. When a black powder, called toner, is put on the zinc oxide plate, it sticks only to the parts of the plate still electrically charged. The black toner is then transferred to another piece of paper, where it is made to stick to the paper by heating it. This reproduces the original document.

Zinc chemically reacts with hydrochloric acid (HCl). The reaction produces zinc chloride, a compound, and hydrogen, a gas. The hydrogen gas is what blows up the balloons in this photograph.

Zinc Oxide and Rubber

Zinc oxide is also used in the vulcanization of rubber. Vulcanization is a chemical treatment that makes rubber stronger and more elastic. This process was discovered in 1839 by American inventor Charles Goodyear (1800–1860). Before that time, natural rubber was not very useful. It was fine at room temperature, but it got brittle and cracked in cold weather. In hot weather, it melted and stuck to everything. Goodyear started experimenting with rubber in the early 1800s. In 1839, while doing an experiment, he accidentally dropped some rubber mixed with sulfur onto a hot stove. The resulting rubber was stronger, harder, more elastic, and held up to the weather better. Zinc oxide is used to make the vulcanizing process happen and to make it go faster.

Zinc Chloride and Batteries

Zinc chloride ($ZnCl_2$) is a white solid chemical compound that is soluble, or dissolves, in water. When it is dissolved, the solution conducts electricity very well. Electrolytes are substances that conduct electricity when they are melted or dissolved in water. Zinc chloride is an electrolyte.

At right is a cross-section of a Leclanché cell produced by Ever Ready in 1957. The central carbon rod and zinc-containing battery wall are visible.

This ability to conduct electricity makes zinc chloride useful in making dry-cell batteries. Dry-cells are the types of batteries found in flashlights, portable radios, and remote-controlled toys. Batteries use a chemical reaction to produce electricity.

More than 100 years ago, a French chemist named Georges Leclanché (1839–1882) invented carbon-zinc batteries. One part of a carbon-zinc battery is a paste of ammonium chloride (NH_4Cl), zinc chloride, and a little water. The negative electrode of the battery is the zinc that makes up the wall of the battery. The positive electrode is a rod of carbon. Unfortunately, carbon-zinc batteries have a few problems. First, they deteriorate quickly in cold weather and do not hold their charge well. Secondly, in order to supply electrons to produce electricity, the zinc wall of the battery turns into zinc ions, and this breaks down the wall, allowing the contents of the battery to leak out.

Some of these problems have been solved by alkaline batteries. The main difference between carbon-zinc and alkaline batteries is alkalines do not contain ammonium chloride (NH_4Cl). Ammonium chloride could

Glowing Zinc Sulfide

Zinc sulfide (ZnS), another zinc compound, is a white or yellow powder. It is also a phosphor. Phosphors give off light when they are hit with electrons, ultraviolet light, or X-rays. So, zinc sulfide is used to coat the insides of television and cathode-ray computer monitor tubes. When a stream of electrons flows through the tubes, it produces the light you see from television and computer screens. Zinc sulfide is also used in glow-in-the-dark toys and decorations and in fluorescent lightbulbs.

release ammonia gas (NH_3), which creates pressure inside the battery, causing it to leak. The ammonium chloride was replaced by potassium hydroxide (KOH), a base, or alkaline material. The name alkaline battery comes from the use of a base in the electrolyte. Alkaline batteries hold their charge better, work better in cold weather, and do not spill their contents as often as the old carbon-zinc batteries did.

Alkaline batteries are not the only batteries that use zinc. Zinc-air and zinc-silver batteries are also used in the electronics industries to power hearing aids, calculators, and wristwatches.

Zinc chloride is hygroscopic, which means it attracts and retains water. It is also an astringent, meaning that it makes body tissues shrink and pull together. These properties make zinc chloride useful in the manufacture of glue, rayon, paper, cosmetics, disinfectants, and firefighting foam. It protects wood from decay and insects, and it kills moss that can form on rooftops, sidewalks, and driveways. It can also be used to separate oil from water.

Zinc oxide, zinc sulfide, and zinc chloride are only a few of the more widely used compounds of zinc. Another zinc compound, zinc phosphide (Zn_3P_2), is a black-grey powder that smells like garlic and is used to control pesky rodents like mice, gophers, rats, and squirrels. When eaten, zinc phosphide reacts with water and acid in the stomach to produce phosphine gas (PH_3), which is toxic. The compound zinc sulfate ($ZnSO_4$) is used as a wood preservative, in fertilizer, and in the manufacture of rayon, a fiber used in clothing.

Chapter Five
Zinc and You

Zinc has countless industrial and commercial uses. It is also a biologically important element. Your body needs zinc to function properly. Fortunately, you can easily get the zinc you need by eating a balanced, healthy diet. Red meat is the best dietary source of zinc. Poultry and seafood also contain a lot of zinc. Vegetarians can get enough zinc in their diet by eating sunflower or pumpkin seeds, brewer's yeast, maple syrup, and bran.

Too much of a good thing can be bad, though. Vomiting, stomach cramps, diarrhea, and fever are all symptoms of eating too much zinc. The recommended dietary allowance of zinc for adult women and young boys and girls is 8 milligrams. Females fourteen to eighteen years old should get 9 mg of zinc per day. Males older than fourteen should get 11 mg of dietary zinc daily.

Zinc and Your Health

Enzymes are proteins involved in important chemical reactions in your body. At least 100 enzymes require tiny amounts of zinc in order to function properly. More than 90 percent of the zinc in the body is found in muscles and bones, but at least a little bit of zinc is present in every cell of the body.

Oysters, pumpkin seeds, and red meat contain the zinc your body needs to function properly. Poultry, beans, nuts, and sardines are also good dietary sources of zinc.

Zinc plays a vital role in the immune system, the body system that protects you from disease and infection. The mineral is necessary for the development and activation of T-lymphocytes, a type of white blood cell that fights infections. Without enough zinc, wounds do not heal as quickly, and the body may not be able to recognize and fight off certain types of infections. Zinc also helps speed up the renewal of skin cells. For this reason, it is a common ingredient in diaper rash ointment.

The mineral is partly responsible for controlling oil-secreting glands, called sebaceous glands, and it is required for making collagen, a protein found in skin and other connective tissue. Zinc has a role in maintaining healthy nails, skin, and hair and preventing dandruff. A zinc deficiency

Do Zinc Lozenges Really Work?

Zinc cough drops cannot prevent someone from getting a cold. They do not exactly cure the common cold either. But some studies have shown that they can reduce the severity of cold symptoms, like runny nose and cough, and shorten the duration of a cold in adults. Scientists do not know exactly why zinc has this effect. They think the zinc may attach to the cold virus and physically prevent it from entering the body's cells. Another hypothesis is zinc could prevent the virus from building proteins needed to reproduce.

Unfortunately, for the lozenges to be effective at all, they must be taken soon after cold symptoms begin and they have to be taken about every two hours. People have to decide for themselves if the bad taste and nausea that often come with taking zinc lozenges are worth having a cold with milder symptoms for a shorter period of time.

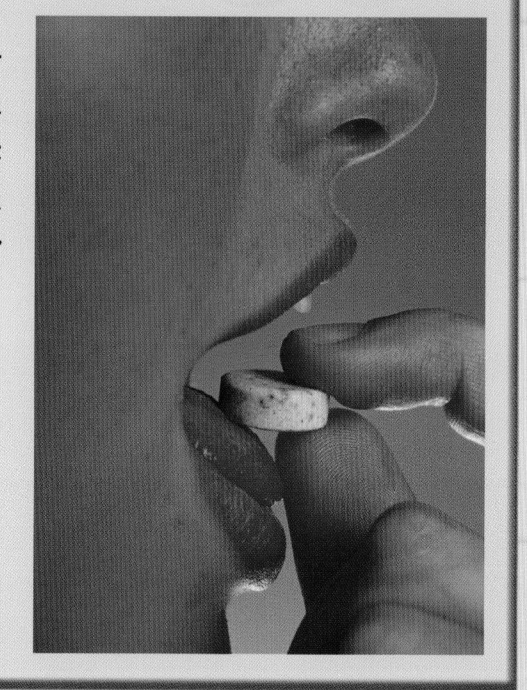

Zinc lozenges may make cold symptoms milder and shorten the amount of time someone has a cold. However, some wonder if the negative side effects are worth it.

can lead to dull, thin-looking hair; hair loss; white spots under the fingernails; or skin problems, like acne.

Zinc plays an important role in appetite and the senses of taste and smell. Both taste buds and olfactory, or smell, cells contain proteins that depend on zinc to function properly. People with problems tasting and smelling their food often lose their appetites. In fact, loss of the sense of smell or appetite is one of the first indications that someone is not getting enough dietary zinc.

High levels of zinc are found in the retina of the eye, too. Scientists have discovered proper zinc concentrations are needed to protect the eyes from a disease called age-related macular degeneration, or AMD. AMD can lead to partial or total loss of sight if it is left untreated. Zinc also helps prevent night blindness and the development of cataracts, a condition that clouds the lens of the eye, reducing vision.

This man suffers from age-related macular degeneration (AMD), a disease that leads to blindness. Dietary zinc supplements can slow down the progress of this and other eye diseases.

Zinc Deficiency

Zinc deficiency is not usually a problem in developed countries like the United States. In developing countries, however, zinc deficiency is the fifth-leading risk factor for developing a disease. Major zinc deficiency can lead to stunted growth and learning, memory problems, and behavioral issues. The most common symptoms of a mild zinc deficiency include dry and rough skin, dull-looking hair, brittle fingernails, white spots on nails, reduced taste and smell, loss of appetite, mood swings, frequent infections, delayed wound healing, and acne.

Zinc in Crops

Zinc is an important nutrient for plants. It is needed during photosynthesis, the process most plants use to convert energy from the sun into food. The element is also needed by plants for growth and protection from disease. When crops are deficient in zinc, the plants are smaller and some of them may die. This can have a major impact on a farmer because fewer, smaller plants mean less money. So, farmers keep a close watch on the amount of zinc their plants get by testing the soil and plant leaves. If they

Zinc deficiency in plants often shows up as yellow spots. Farmers can protect their crops by adding fertilizers that contain zinc.

are not getting enough zinc, the farmer might add to the natural amount of zinc in the soil with a fertilizer that has zinc in it.

Summing Up Zinc

Now you can see why zinc is called the Great Protector. It can prevent our skin from getting sunburned, it protects our buildings and bridges, and it boosts our immune system. It protects both humans and plants from disease and helps us grow strong and healthy. Zinc's compounds light up our world with fireworks, glow-in-the-dark watch dials, and batteries that produce electricity for our flashlights and other electronic equipment. Zinc allows us to make photocopies and tough rubber tires for our cars, and to color our world with white and yellow paint pigments. No matter how you look at it, zinc is an essential part of our lives.

The Periodic Table of Elements

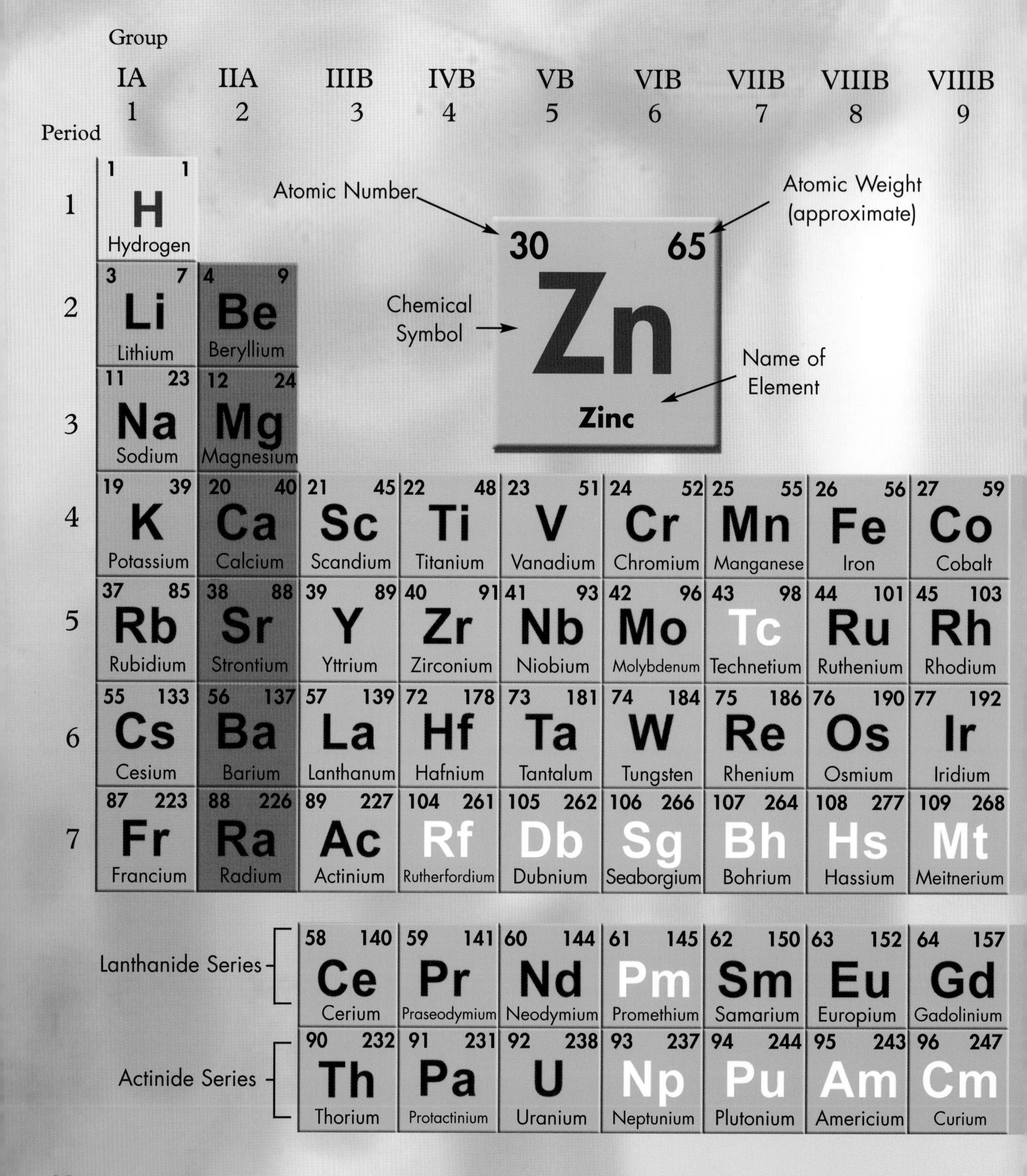

Period	IA 1	IIA 2	IIIB 3	IVB 4	VB 5	VIB 6	VIIB 7	VIIIB 8	VIIIB 9
1	1 1 H Hydrogen								
2	3 7 Li Lithium	4 9 Be Beryllium							
3	11 23 Na Sodium	12 24 Mg Magnesium							
4	19 39 K Potassium	20 40 Ca Calcium	21 45 Sc Scandium	22 48 Ti Titanium	23 51 V Vanadium	24 52 Cr Chromium	25 55 Mn Manganese	26 56 Fe Iron	27 59 Co Cobalt
5	37 85 Rb Rubidium	38 88 Sr Strontium	39 89 Y Yttrium	40 91 Zr Zirconium	41 93 Nb Niobium	42 96 Mo Molybdenum	43 98 Tc Technetium	44 101 Ru Ruthenium	45 103 Rh Rhodium
6	55 133 Cs Cesium	56 137 Ba Barium	57 139 La Lanthanum	72 178 Hf Hafnium	73 181 Ta Tantalum	74 184 W Tungsten	75 186 Re Rhenium	76 190 Os Osmium	77 192 Ir Iridium
7	87 223 Fr Francium	88 226 Ra Radium	89 227 Ac Actinium	104 261 Rf Rutherfordium	105 262 Db Dubnium	106 266 Sg Seaborgium	107 264 Bh Bohrium	108 277 Hs Hassium	109 268 Mt Meitnerium

Series							
Lanthanide Series	58 140 Ce Cerium	59 141 Pr Praseodymium	60 144 Nd Neodymium	61 145 Pm Promethium	62 150 Sm Samarium	63 152 Eu Europium	64 157 Gd Gadolinium
Actinide Series	90 232 Th Thorium	91 231 Pa Protactinium	92 238 U Uranium	93 237 Np Neptunium	94 244 Pu Plutonium	95 243 Am Americium	96 247 Cm Curium

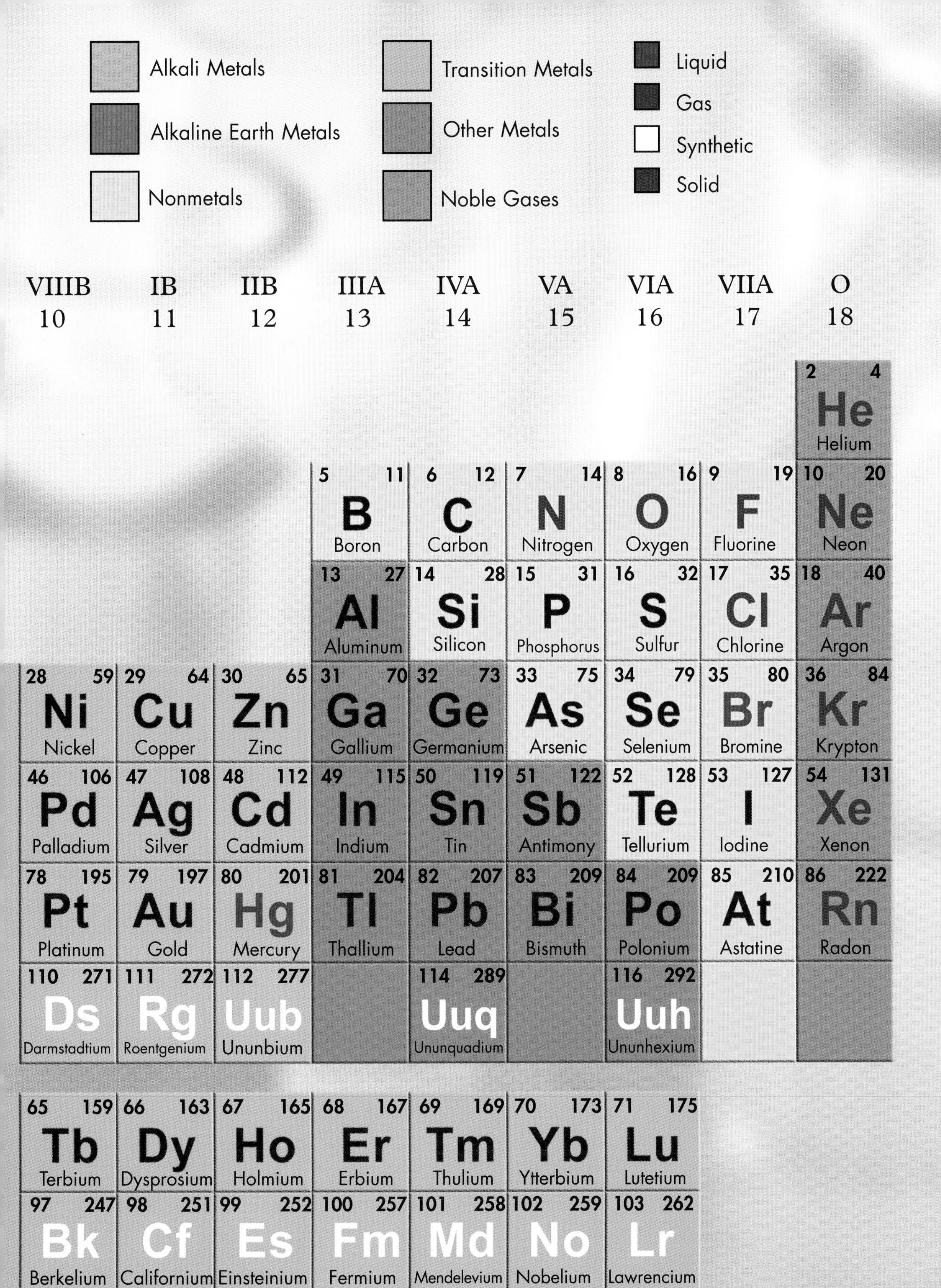

Alkali Metals
Alkaline Earth Metals
Nonmetals
Transition Metals
Other Metals
Noble Gases
Liquid
Gas
Synthetic
Solid
VIIIB 10
IB 11
IIB 12
IIIA 13
IVA 14
VA 15
VIA 16
VIIA 17
O 18
2 4 He Helium
5 11 B Boron
6 12 C Carbon
7 14 N Nitrogen
8 16 O Oxygen
9 19 F Fluorine
10 20 Ne Neon
13 27 Al Aluminum
14 28 Si Silicon
15 31 P Phosphorus
16 32 S Sulfur
17 35 Cl Chlorine
18 40 Ar Argon
28 59 Ni Nickel
29 64 Cu Copper
30 65 Zn Zinc
31 70 Ga Gallium
32 73 Ge Germanium
33 75 As Arsenic
34 79 Se Selenium
35 80 Br Bromine
36 84 Kr Krypton
46 106 Pd Palladium
47 108 Ag Silver
48 112 Cd Cadmium
49 115 In Indium
50 119 Sn Tin
51 122 Sb Antimony
52 128 Te Tellurium
53 127 I Iodine
54 131 Xe Xenon
78 195 Pt Platinum
79 197 Au Gold
80 201 Hg Mercury
81 204 Tl Thallium
82 207 Pb Lead
83 209 Bi Bismuth
84 209 Po Polonium
85 210 At Astatine
86 222 Rn Radon
110 271 Ds Darmstadtium
111 272 Rg Roentgenium
112 277 Uub Ununbium
114 289 Uuq Ununquadium
116 292 Uuh Ununhexium
65 159 Tb Terbium
66 163 Dy Dysprosium
67 165 Ho Holmium
68 167 Er Erbium
69 169 Tm Thulium
70 173 Yb Ytterbium
71 175 Lu Lutetium
97 247 Bk Berkelium
98 251 Cf Californium
99 252 Es Einsteinium
100 257 Fm Fermium
101 258 Md Mendelevium
102 259 No Nobelium
103 262 Lr Lawrencium

Glossary

alloy Combination of two or more metals.

anode The electrode at which electrons are removed from a substance, where a substance is oxidized.

atom Basic building block of all matter.

cathode The electrode at which electrons are added to a substance, where a substance is reduced.

compound Substance formed when two or more elements chemically react and form chemical bonds.

ductile Capable of being stretched into wires without breaking.

electrolyte Substance that, when in solution or melted, breaks down into ions and conducts an electric current.

electron Negatively charged subatomic particle found in energy levels outside an atom's nucleus.

electron configuration Arrangement of electrons in energy levels around the nucleus of an atom.

element Substance made up of only one type of atom.

enzyme Protein that speeds up chemical reactions in a living organism.

ion Electrically charged particle produced when an atom or a group of atoms gains or loses electrons.

ionic bond Chemical bond formed between oppositely charged ions. This is the type of bond that forms between a metal and a nonmetal.

isotope Any of two or more forms of an element that contain the same number of protons but have different numbers of neutrons in the nucleus.

malleable Capable of being hammered, rolled, pressed, or molded into different shapes without breaking.

neutron Subatomic particle with no charge found inside an atom's nucleus that, along with the atom's protons, makes up most of the mass of the atom.

ore Mineral deposit containing a metal that is valuable enough to be mined.

oxidation Addition of oxygen to, or the removal of electrons from, an element or compound.

proton Positively charged subatomic particle found inside an atom's nucleus that, along with the atom's neutrons, makes up most of the mass of the atom.

reduction Loss of oxygen from, or the addition of electrons to, an element or compound.

salt Type of ionic compound.

soluble Able to be dissolved.

transition metals Metals in periodic table groups 3 through 12 (IB through VIIIB).

valence electrons Electrons in the outermost energy level of an atom. These are the electrons involved in chemical bonding.

For More Information

American Chemical Society
1155 Sixteenth Street NW
Washington, DC 20036
(800) 227-5558 (U.S. only)
(202) 872-4600 (outside the U.S.)
E-mail: help@acs.org
Web site: http://www.chemistry.org/kids
The American Chemical Society Web site contains a collection of hands-on activities, games, interactive activities, and articles written for children. Print materials also can be obtained from their online store.

American Galvanizers Association
6881 South Holly Circle, Suite 108
Centennial, CO 80112
(800) 468-7732
E-mail: aga@galvanizeit.org
Web site: http://www.galvanizeit.org
The American Galvanizers Association Web site has fast facts about zinc and hot-dip galvanization.

American Zinc Association (AZA)
2025 M Street NW, Suite 800
Washington, DC 20036
(202) 367-1151
E-mail: zincinfo@zinc.org
Web site: http://www.zinc.org

The American Zinc Association seeks to educate the public about the metal. The AZA publishes a wide range of documents regarding zinc, many of which can be downloaded from its Web site.

Canadian Zinc Corporation
Suite 4, Government of Canada Building
9606 – 100 Street
Fort Simpson, NT X0E 0N0
Canada
(867) 695-3963
Web site: http://www.canadianzinc.com
Canadian Zinc Corporation mines zinc, copper, lead, and silver in the Mackenzie Mountains of the Northwest Territories. Information about the geology of the mine and a photo gallery can be found on its Web site.

Web Sites

Due to the changing nature of Internet links, Rosen Publishing has developed an online list of Web sites related to the subject of this book. This site is updated regularly. Please use this link to access the list:

http://www.rosenlinks.com/uept/zinc

For Further Reading

Baldwin, Carol. *Compounds, Mixtures, and Solutions* (Material Matters/Freestyle Express). Chicago, IL: Raintree, 2004.

Baldwin, Carol. *Metals* (Material Matters/Freestyle Express). Chicago, IL: Raintree, 2005.

Gray, Leon. *Zinc* (The Elements). New York, NY: Marshall Cavendish Benchmark, 2006.

Kjelle, Marylou Morano. *The Properties of Metals*. New York, NY: Rosen Publishing, 2006.

Kjelle, Marylou Morano. *The Properties of Salts*. New York, NY: Rosen Publishing, 2006.

Ricciuti, Edward. *Rocks and Minerals.* New York, NY: Scholastic Reference, 2002.

Stille, Darlene. *Chemical Change: From Fireworks to Rust.* Minneapolis, MN: Compass Point Books, 2005.

Tocci, Salvatore. *The Periodic Table*. New York, NY: Children's Press, 2004.

Tocci, Salvatore. *Zinc* (A True Book). New York, NY: Children's Press, 2005.

Watt, Susan. *Mercury.* New York, NY: Marshall Cavendish Benchmark, 2004.

Bibliography

American Galvanizers Association. "Hot Dip Galvanizing for Corrosion Protection of Steel Products." 2000. Retrieved June 8, 2007 (http://www.gordtelecom.com/PDF-Files/Hot%20Dip%20Galvanizing,%20A%20guide%20to.pdf).

Americans for Common Cents. "The History of the United States Penny in the 20th Century." July 2006. Retrieved June 7, 2007 (http://www.pennies.org/history/eight.html).

American Zinc Association. "Zinc World." September 2006. Retrieved June 7, 2007 (http://www.zinc.org).

Booras, Charles. "SPF, UVB and UVA Protection Explained." May 1998. Retrieved June 8, 2007 (http://www.jaxmed.com/articles/wellness/spf.htm).

Calvert, J. B. University of Denver. "Zinc and Cadmium." November 2002. Retrieved June 8, 2007 (http://www.du.edu/~jcalvert/phys/zinc.htm#Mine).

Corrosion Doctors. "Georges Leclanché." May 2007. Retrieved June 8, 2007 (http://www.corrosion-doctors.org/Biographies/LeclancheBio.htm).

Emsley, John. *Nature's Building Blocks: An A–Z Guide to the Elements.* New York, NY: Oxford University Press, 2001.

Georgia State University: Department of Physics and Astronomy. "Batteries." December 2001. Retrieved June 8, 2007 (http://hyperphysics.phy-astr.gsu.edu/hbase/electric/battery.html#c3).

Johns Hopkins University. "Lava Lamp-Like Process Caused World's Largest Zinc Deposit." April 2005. Retrieved June 7, 2007 (http://www.jhu.edu/~news_info/news/home05/apr05/zinc.html).

Leach, David. "The Giant Red Dog Massive Sulfide Deposit Brooks Range, Alaska." U.S. Geological Survey. Retrieved June 7, 2007 (http://www.cpgg.ufba.br/metalogenese/Leach1.htm).

Madison Industries, Inc. "Zinc Chloride." February 2007. Retrieved June 7, 2007 (http://www.oldbridgechem.com/ZnCl2.html).

Madison Industries, Inc. "Zinc Sulfate Solution." February 2007. Retrieved June 7, 2007 (http://www.oldbridgechem.com/ZnCl2.html).

MedicineNet, Inc. "Calamine Lotion." March 2005. Retrieved June 8, 2007 (http://www.medicinenet.com/calamine_lotion-topical/article.htm).

MedicineNet, Inc. "Zinc Lozenges as a Cold Remedy." January 2007. Retrieved June 7, 2007 (http://www.medicinenet.com/zinc_lozenges_as_a_cold_remedy/page2.htm).

Mineral Information Institute. "Zinc." October 2004. Retrieved June 7, 2007 (http://www.mii.org/Minerals/photozinc.html).

National Electrical Manufacturers Association. "Button Cell Collection." January 2003. Retrieved June 7, 2007 (http://www.nema.org/gov/ehs/committees/drybat/upload/Buttoncellcollection.pdf).

National Institutes of Health: Office of Dietary Supplements. "Facts About Zinc." December 2002. Retrieved June 7, 2007 (http://dietary-supplements.info.nih.gov/factsheets/cc/zinc.html#what).

National Inventors Hall of Fame. "Charles Goodyear." 2002. Retrieved June 8, 2007 (http://www.invent.org/hall_of_fame/68.html).

Thomas Jefferson National Accelerator Facility, Office of Science Education. "It's Elemental—the Element Zinc." Retrieved June 7, 2007 (http://education.jlab.org/itselemental/ele030.html).

University of Illinois Urbana-Champaign: Department of Materials Science and Engineering. "Metals: Scientific Principles." March 2000. Retrieved June 8, 2007 (http://matse1.mse.uiuc.edu/~tw/metals/prin.html).

U.S. Geological Survey. "USGS Minerals Information: Zinc." May 2007. Retrieved June 7, 2007 (http://minerals.usgs.gov/minerals/pubs/commodity/zinc).

Index

About the Author

Kristi Lew is a professional K–12 educational writer with degrees in biochemistry and genetics. A former high school science teacher, she specializes in writing textbooks, magazine articles, and nonfiction books about science, health, and the environment for students and teachers.

Photo Credits

Cover, pp. 1, 13, 16, 17, 20, 22, 38–39 Tahara Anderson; p. 5 © Warren Jacobi/Corbis; p. 7 Wikimedia Commons; pp. 8, 35 © Getty Images; p. 9 © Ludo Kuipers/Corbis; p. 10 © Ben Johnson/Photo Researchers, Inc.; pp. 24, 29 © SSPL/The Image Works; p. 27 © Kevin Fleming/Corbis; p. 28 © Charles D. Winters/Photo Researchers, Inc.; p. 33 Shutterstock.com; p. 34 © Will & Deni McIntyre/Photo Researchers, Inc.; p. 36 © Nigel Cattlin/Visuals Unlimited.

Designer: Tahara Anderson; **Editor:** Christopher Roberts